# EXPLORER
# MAMMALS!

By **Nick Forshaw** & **William Exley**

**What on Earth Books**

What on Earth Publishing Ltd., The Black Barn, Wickhurst Farm, Tonbridge, Kent TN11 8PS, United Kingdom

First published in 2019.

Produced in association with the Natural History Museum, London.

Written by Nick Forshaw.
Illustrated by William Exley.
Designed by Andy Forshaw and Assunção Sampayo.
Edited by Patrick Skipworth.

ISBN: 978-0-9955766-2-9

Printed by Waiman in China.

10 9 8 7 6 5 4 3 2 1

whatonearthbooks.com

# CONTENTS

**MISSION INSTRUCTIONS**

**AGENT ASSIGNED:** Agent Osprey

**YOUR MISSION:** Calling Agent Osprey! You must once again venture into the unknown. Explorer HQ need a report on mammals ASAP! Go out and examine every furry critter you can find.

CLASSIFIED

*AGENT OSPREY*
*M p Archives*

I'm Ada Osprey, Senior Librarian of the Eagle-Eyed Explorer Club. Years ago, the club recruited me as one of its special agents — code name, Agent Osprey.

Most days you'll find me in the basement of the library, organising the atlases. But I always keep my rucksack at the ready, in case the club needs me for one of its urgent missions. I never know when to expect a call.

**STAFF ROOM** ←

**LIBRARY** ←

**MAMMALS!** →

*Mission Checklist*

1x Explorer Field Notebook
1x Magnifying Glass
1x Binoculars
1x Agent Issue Torch
1x Portable Time Machine
1x Sandwich (cheese & pickle)

My latest mission is not only top secret, but very dangerous, so keep it hush-hush. The club has ordered me to file a report on the entire history of mammals as soon as possible. It looks like I'll need to go back millions of years in time.

I'll need to fill out a Journal to keep track of my discoveries. I'm also drawing up a Timeline to show the life story of mammals, from their ancient ancestors to the present day.

No time to waste. Are you feeling brave? Let's go!

# 1. WHAT ARE MAMMALS?

*DIMETRODON*

'What does a wolf have to do with a walrus? What's the link between a monkey and a mouse? How can a cheetah be related to a bottlenose dolphin? It's our job to find out the answers to all these questions, as we explore ... mammals!'

## Scaly ancestors

The ancient ancestors of modern mammals didn't look much like the furry creatures which roam the Earth today. 280 million years ago, huge *Dimetrodon* plodded through the undergrowth on the hunt for prey. Giant sails waggled on their backs, as they chased down reptiles to eat. *Dimetrodon* were enormous lizard-like creatures. They could grow over 3.5m-long, with scaly skin and snouts full of sharp teeth. They were part of an animal group called synapsids, which includes the ancestors of the mammals we know today. Long before the dinosaurs existed, these synapsids were Earth's top land predators.

The ancestors of mammals survived the extinction

But then came the worst extinction the world has ever seen. Extinction is when species die out completely. During this period, the climate began to change rapidly. Nine out of 10 species were wiped out. Lots of synapsids went extinct, but one group clung on through the disaster. They were much smaller than hefty predators like *Dimetrodon* and some of them even had fur and whiskers. Over time these animals developed into the first mammals, around 200 million years ago. Today, mammals can be found deep in the oceans and on every continent except Antarctica.

## All in order?

But what are mammals? Where do they live and how do they survive? Mammals live in all sorts of different habitats, on dry land, in the sea, even in the air. Some have shaggy fur, others have smooth skin. Some eat meat, some eat plants, some will eat anything! To make sense of this variety, scientists put mammals into special groups.

Many meat-eating mammals are put in a big group called 'Carnivora'. To make things even clearer, scientists put different members of Carnivora into smaller groups. One type of a smaller group is a 'genus'. *Canis* is the genus which includes wolves and dogs, for example. But not all wolves and dogs are the same either, so scientists break the groups down again, into 'species'. Creatures in a species all share similar characteristics. One species can't breed with another. *Canis lupus* is the species name for the grey wolf. 'Canis' tells us this creature is a wolf or a dog, and 'lupus' tells us precisely which species.

*Mammals have evolved into lots of different species*

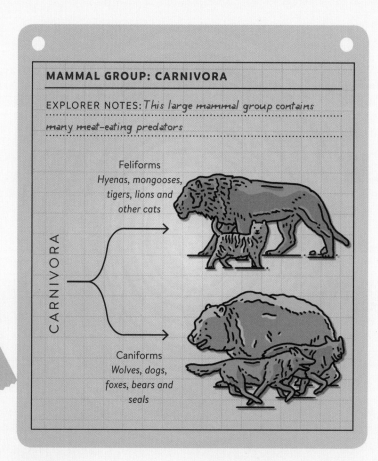

**MAMMAL GROUP: CARNIVORA**

EXPLORER NOTES: *This large mammal group contains many meat-eating predators*

CARNIVORA

Feliforms
*Hyenas, mongooses, tigers, lions and other cats*

Caniforms
*Wolves, dogs, foxes, bears and seals*

## Evolution

But how are mammals related to scaly *Dimetrodon* in the first place? 'Evolution' is the process where all living things change over thousands of years. These changes happen slowly, when a new generation is born. The babies may have more fur than their parents, for example, or sharper claws. Some changes can help them survive. When they reproduce, their own babies may have the same features. These features are called 'adaptations'. Warm fur, sharp teeth and keen eyesight are just a few adaptations that may help mammals to survive. Over thousands of years, entirely new species can evolve. This is how completely new types of animal might develop from creatures like *Dimetrodon*.

*'Turn the page to discover more about what makes mammals special!'*

### Egg-ception to the rule

Whether birds or lizards, insects or fish, almost all animals lay eggs from which their young will hatch. But mammals are different. The vast majority of mammals don't lay eggs. Instead, they give birth to live young. Their babies develop safely inside their mother's womb until their mothers are ready to give birth. As soon as the infants are born, they suckle on the nutritious milk their mothers produce in their mammary glands.

PLATYPUS ARE MAMMALS WHICH LAY EGGS

### Fur coats

Most mammals have fur coats that keep them warm. These can be short and soft like a rabbit's or a jaguar's, or long and shaggy like a sheep's or an orangutan's. Even smooth-skinned whales have a few bristly hairs when they are young. The patterned fur of giraffes and jaguars helps them hide in their habitat. This is called camouflage. It keeps them safe from predators or unseen by prey.

## Natural warmth

Nearly all mammals are warm-blooded – but what does that really mean? Warm-blooded animals have bodies that are naturally warm. They don't need to lie in the sun to warm up, like cold-blooded creatures such as lizards or snakes. Fur helps to trap the body's heat. Not all mammals have fur however. Even though naked mole rats are warm-blooded too, they huddle together underground to stay warm when it gets too cold.

FATTY BLUBBER KEEPS WHALES WARM

## Unusual bones

Some scientists like to get right down to the bone when exploring wildlife. Looking at a mammal's skeleton, scientists can see how they have jaws and teeth which let them chew up their food to digest it better. Also, their legs do not sprawl out like a lizard's. This helps many of them to walk easily and run long distances. Mammals even have tiny bones inside their ears. These bones help to give them very sensitive hearing.

# 2. EARLY MAMMALS

HADROCODIUM

'In the Jurassic period, 200 million years ago, the dinosaurs ruled over the Earth. They prowled through trees and ferns on the hunt for food. But hidden out of sight, other creatures had evolved too. Darting out between the dinosaurs' toes were the first mammals! Let's take a closer look!'

## New arrivals

Deep in the undergrowth, the first mammals were scurrying from one hiding place to another. Some of these early mammals were tiny, no bigger than your finger. They had furry bodies, with long tails that wriggled behind them as they scuffled about. If they heard the stomping footsteps of a dinosaur, they dashed under a rock for cover.

One of the earliest known mammals is *Hadrocodium*. It was so small it weighed less than an acorn. It had a furry body and tail, and a short nose. *Hadrocodium* was always on the lookout for tiny insects to eat. It could disappear in

Many early mammals ate insects

an instant if a larger creature approached. Even though it was small, *Hadrocodium* had a big brain compared to many other animals at the time. Scientists think this means it had well-developed senses. Like a mouse, it could smell and hear very well. It may have been able to hear danger approaching from far away so it had time to find a safe hiding spot. And its nose could sniff out a bug hiding beneath the leaves. Over millions of years, early mammals like mouse-sized *Hadrocodium* evolved in new forms, developing into all the mammals alive on Earth today.

## Small survivors

But if there were so many dinosaurs prowling around, how did the miniature mammals survive? Early mammals' small size was one secret to their survival. It helped them avoid becoming a dinosaur's next meal. Some mammals were so small, larger dinosaurs may not have even noticed them. If a hungry dinosaur appeared out of the bushes, these mammals could quickly disappear into the roots and brambles.

Small bodies may have helped them in other ways too. Large creatures need larger meals to survive. Big dinosaurs could have struggled when food was scarce. But the smaller mammals, many of which ate insects, were able to keep going.

Mammal fossils have been found at dinosaur nesting sites

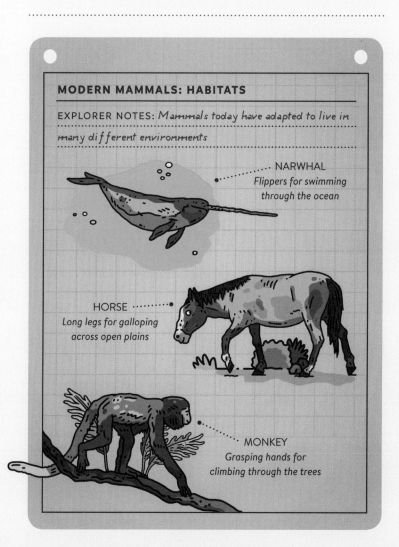

**MODERN MAMMALS: HABITATS**

EXPLORER NOTES: *Mammals today have adapted to live in many different environments*

**NARWHAL**
*Flippers for swimming through the ocean*

**HORSE**
*Long legs for galloping across open plains*

**MONKEY**
*Grasping hands for climbing through the trees*

## At home in your habitat

An animal's habitat is the part of its environment that it calls home. It's the place where it finds the food and shelter to survive. The Jurassic forest was the habitat of many early mammals. They sheltered beneath the vegetation and lots of them fed on the bugs that lived in the trees. Among the ferns and bushes, they were safe from prowling predators.

Species evolve in ways that help them thrive in parts of the habitat others haven't found yet. Or their adaptations might help them push out a species which had already found success in a habitat. This is how species find their own 'niche', a place where they fit in. They can feed on food that other creatures are not already eating, or live in spaces other creatures have not yet taken over.

*'Turn the page to discover more about the first mammals!'*

## What were some of the first mammals?

Giant dinosaurs dominated the natural world in the Jurassic period. But down in the undergrowth, mammals had their own ways of surviving too. Some foraged among roots and ferns. Others dug up termite mounds. Some species swam in rivers or even sailed from tree to tree. Here are my field notes on some of the most curious prehistoric mammals …

### CASTOROCAUDA | LENGTH – 42cm

EXPLORER NOTES: *Swift swimming with a flat tail*

Paddling through pond plants and weeds, *Castorocauda* hunted for fish and frogs. Its broad, flat tail helped it to swim swiftly through the water. Webbed paws allowed it to twist and turn to catch its wriggling prey.

LIVED *164 million years ago*

### VOLATICOTHERIUM | LENGTH – 25cm

EXPLORER NOTES: *Furry mammal that glided from tree to tree*

Today, bats are the only mammals to have evolved the power to fly. This early mammal was almost there – it could glide! High up in a tree, *Volaticotherium* may have spotted a bug crawling on the ground below. Springing into the air, it would spread out the skin between its front and back legs like a parachute and swoop down onto its prey. Life in the treetops kept it safe from predators on the ground. But it could still glide between the trees when it needed to make a quick getaway.

LIVED *164 million years ago*

### FRUITAFOSSOR | LENGTH – 15cm

EXPLORER NOTES: *Small but strong mammal that broke into insect nests*

Despite its small size, this mammal was a powerful creature with a compact body and tough, blunt teeth. *Fruitafossor* used its muscly arms to break open termite mounds. Then it would stick its snout deep into the nest and feed on the bugs and grubs hidden inside.

LIVED *160 million years ago*

## RUGOSODON | LENGTH – 17cm

EXPLORER NOTES: *Nimble mammal with a varied diet*

This slim mammal had a short tail and nimble body. It was an (omnivore,) meaning it could eat lots of different things. Bugs and plants were both on the menu. Its teeth allowed it to crack into seeds or gobble up insects hiding among the leaves.

**LIVED** *160 million years ago*

## CRONOPIO | LENGTH – 10cm

EXPLORER NOTES: *Tiny mammal with humongous sharp teeth*

A small, furry mammal with (gigantic canine teeth.) Scientists even nicknamed this species the 'sabre-toothed squirrel'. Its sharp gnashers and thin snout would have helped it catch all kinds of different insects.

**LIVED** *100 million years ago*

## REPENOMAMUS | LENGTH – 70cm–1m

EXPLORER NOTES: *Stocky mammal that hunted down dinosaurs*

Not all early mammals were tiny. This chunky (hunter) lived alongside the dinosaurs. It was a meat-eater with sharp pointed teeth. Most mammals would have scurried away if they heard a dinosaur approaching. But this predator did the opposite. It chased after young dinosaurs and caught them as prey!

**LIVED** *130 million years ago*

# 3. ANCIENT MAMMALS

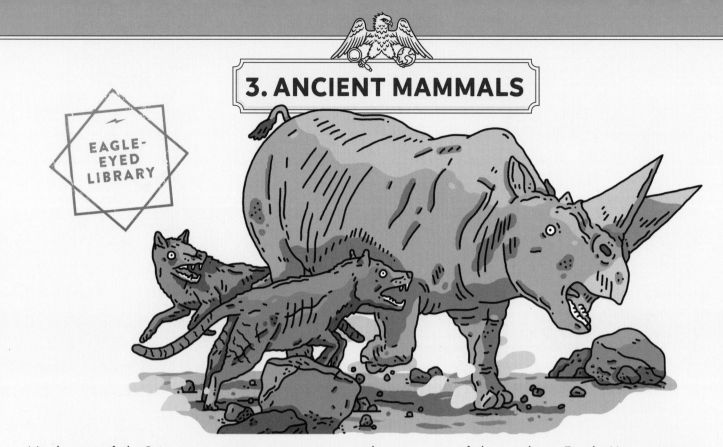

'At the end of the Cretaceous period, dinosaurs were the most powerful animals on Earth. Huge meat-eaters chased after prey. Towering giants fed on leaves from the tallest trees. But down on the ground, early mammals scampered through the undergrowth. Then, 66 million years ago, something happened that changed everything. And the mammals emerged from their burrows into a brand new world.'

## A new world

The Cretaceous jungle flourished with giant dinosaurs and scurrying mammals. But suddenly, 66 million years ago, the whole world turned upside down. The ground shook, temperatures plunged and huge waves flooded the coasts. Dust and debris were thrown up into the sky so the sun could no longer shine through. Daylight disappeared, causing years of darkness. Life on Earth struggled to survive. What had caused this disaster?

A meteor from outer space had collided with the planet. The impact was so powerful the whole world shook. Clouds of dust billowed into the air and settled on every surface. Strong winds tore trees from the ground. The dinosaurs were not able to withstand the disaster. Species by species, they went extinct. But among the rubble, mammals scampered into spaces where they were protected from the chaos. Their furry bodies kept them warm against the cold. Small sizes meant they did not need as much food as the gigantic dinosaurs. Many lived in burrows, safe underground. As the Earth settled into a new phase, the mammals came out of their hiding spaces. And, in a world free from dinosaurs, the mammals took over.

The meteor contributed to global mass-extinctions

**MODERN MAMMALS: PLANT-EATERS**

EXPLORER NOTES: *Plants are the main food source for many mammals*

GIRAFFE
*Browses on leaves in tall trees*

REINDEER
*Grazes on grass and shoots by frozen rivers*

PIG
*Searches with its snout for nuts and fungi*

## Grass-grazers

Slowly, over millions of years, the barren terrain transformed. Grasses spread out over the open plains. With these new forms of plant life, mammals evolved into new forms too. One example was *Mesohippus*. Huge herds of these horse-like mammals crossed the plains. They covered great distances searching for fresh fields to graze on.

Mammals like wildebeest or zebra behave in the same way today. They travel in massive numbers across the African savannah to seek out fresh pastures. In the far northern hemisphere, herds of reindeer or caribou plod steadily through the snow. When they come across a zone where the ice has melted, they stop to graze on the exposed grass.

## Deadly hunters

In the forests and jungles today, powerful tigers hunt for prey in the dark green undergrowth. They spy on their victims before springing out to pounce. But in the prehistoric past, even more deadly mammals were on the prowl. *Smilodon*s were gigantic sabre-toothed cats, with canine teeth up to 30cm long. They wrestled down bison or leapt onto armadillos. Sharp claws helped them pull their victims to the ground. Then they tore into their prey with their lethal fangs.

*Smilodon used its teeth to puncture the necks of its prey*

*'Turn the page to discover some of the most incredible ancient mammals!'*

## Paraceratherium

This plant-eater was the largest mammal ever to live on land! It was eight-metres tall when it stretched up to feed on tall trees. That's about the same height as five humans! Like giraffes today, it could use its rubbery lips to strip leaves from twigs. It had tough, leathery skin, like its cousin today, the rhinoceroses.

## Megatherium

An enormous ground sloth, six-metres tall. This plant-eater had thick fur and a fuzzy snout. It could walk on two legs while it searched for vegetation to tuck into. But if a band of hungry humans came into view, this giant might have made off on all fours. Early humans ate *Megatherium*, although they probably scavenged their corpses. Even so, human hunting still might be one of the reason this species died out 10,000 years ago.

## Mammuthus

Mighty *Mammuthus* – commonly known as the woolly mammoth – stood strong against the whipping winds of the icy Arctic tundra. This ancient mammal had a shaggy coat to keep it warm. It used its long trunk and tusks to pull up plants beneath the snow for food. Early humans were curious about the 4-metre tall mammoth. They painted images of these giant animals on cave walls. But hunting by humans, as well as climate change, eventually contributed to their extinction as well.

## Thylacosmilus

This cat-like predator hunted in South America's forests. But, despite its appearance, it was more closely related to kangaroos than cats! It had enormous canine teeth on its upper jaw. These tucked away neatly into its lower lip. Leaping onto its prey, it would sink its dagger-like fangs into their bodies. Many big cats, like African leopards, hunt in a similar way today.

## Arctotherium

Growing to 3.5 metres long, this ancient bear is the largest ever meat-eating land mammal. It fed on anything it could find, and would spring onto another predator's kill and scare or fight them off when it could. If it couldn't find any meat to feed on, it gobbled up fruit and nuts from bushes. Its massive appetite meant that when food was scarce, it didn't survive long.

# 4. MAMMAL HOMES

*POLAR BEARS & WALRUS*

'As time has passed, environments on Earth have continued to change. Mammals have kept changing and developing too. Today, mammals come in all shapes and forms. They can be found in almost every environment on Earth. In frozen tundras, mammals rely on their fur to protect them from the cold. Even in the dry desert, long-legged mammals survive without much water to drink. And deep in the blue ocean, the largest ever mammals sing to each other under the waves.'

## Frozen snow and desert sand

In the freezing cold of the Arctic tundra, snow storms blast over the icy terrain. This frozen environment is so harsh that not even a single tree can grow. But some mammals have found ways to survive up here in the frozen north. Polar bears' white fur helps them blend in with the ice. And their powerful paws help them catch prey in the snow. They will even attack gigantic walruses, huge blubbery mammals with sharp, pointy tusks. Polar bears keep their cubs in burrows dug beneath the snow, trying to stay warm while their mother searches for food across the ice.

*Camels have adapted to dry desert conditions*

Other mammals survive in completely different environments. In the heat of the North African desert, a train of camels plods steadily over the sand. Camels are mammals with tousled brown fur and humps on their backs. Their long legs help them roam the empty dunes, feeding on whatever vegetation they can find. They eat the tough stems and dried-out leaves of desert plants. Any extra food is stored in their humps. When a camel finds water, it drinks such huge quantities that other animals would collapse. But, for a camel, this may be the only time it has a drink in months.

## World of echo

As daylight turns to dusk, a shadow tumbles through the purple evening sky. The shadow is a bat, on the hunt for insects in the twilight. European bats are small, with soft, furry bodies. They use their leathery wings to fly with perfect precision. They twist and turn and snap up as many flying insects as they can find.

Most bats have poor eyesight, but they use their hearing to figure out what's around them instead. As they fly about, they emit high-pitched squeaks that echo back from any surface – trees, walls, even the bodies of insects. Their highly sensitive ears pick up the echoes, which they use to build an image of their surroundings. Even when the sun has faded, the image lets the bats 'see' the world around them. They fly through the dark but still 'see' the insects they like to eat.

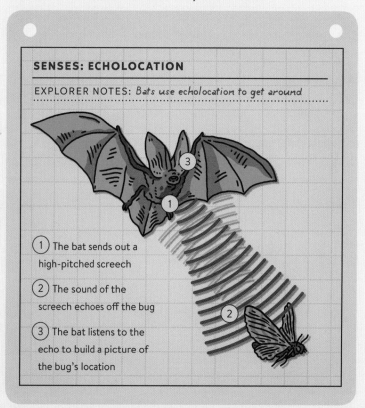

**SENSES: ECHOLOCATION**

EXPLORER NOTES: *Bats use echolocation to get around*

1. The bat sends out a high-pitched screech
2. The sound of the screech echoes off the bug
3. The bat listens to the echo to build a picture of the bug's location

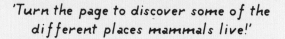

*'Turn the page to discover some of the different places mammals live!'*

*Dolphins are very sociable animals*

## Underwater mammals

Out in the salty seas, dolphins swim steadily through the water. They surface to breathe through their blowholes, before dipping back beneath the waves. Dolphins are highly intelligent marine mammals that live together in small groups called 'pods'. They click and whistle underwater to communicate with each other. They can even hear when another dolphin is calling their name. Dolphin mothers teach their infant 'calves' how to crack open shellfish for food. Their infants can pass on this knowledge to their own offspring when they grow up.

Dolphins are not the only underwater mammals however. Blue whales are the largest mammals ever to live on Earth. They swim in the depths of freezing cold waters, hunting for clouds of krill to feed on. Krill are tiny sea creatures and are the whales' main source of food. As the blue whales travel through the dark oceans, they make noises that sound like songs. The songs are perhaps the whales' way of communicating. They let other whales know they're there.

## Where do mammals make their homes?

Mammals adapt to their environments in different ways. Some hunt secretly in snowy mountains. Others swing through rainforest trees. Some build their own tree houses. Others dig deep tunnels underground. Here are some field notes on where different mammals can be found …

### SNOW LEOPARD

EXPLORER NOTES: *Hides away in rocky dens on lonely mountains*

As night falls over the Himalayan mountains, a snow leopard prowls silently through the dark. Its soft grey fur keeps it warm in the icy air. These muscular cats hunt at night. Their strong paws and sharp teeth help them catch goats or sheep. Snow leopard mothers hide their infants in secret spaces between the rocks until the cubs are ready to venture outside. When the time is right, the cubs trot after their mothers, who teach them how to survive in the snowy wilderness.

### SPIDER MONKEY

EXPLORER NOTES: *Uses its strong tail to leap between trees*

In the steamy rainforests of South America, spider monkeys howl and bark as they bounce from tree to tree. They use their powerful tails to help them swing through the leafy branches. Their tails are so strong and flexible they are like an extra arm. They use their long fingers to pluck bugs crawling on the tree bark, or snaffle up fruits they find between the leaves. If they spot a jaguar prowling down below, they give out a warning whoop. Then they catapult off into the jungle, safe for another day.

### GORILLA

EXPLORER NOTES: *Builds a treetop nest each night to sleep in*

Out in the African jungle, among leaves and ferns, a gorilla keeps watch. Its curious eyes peek out from under a low brow. Gorillas are powerful apes with sturdy bodies covered in black hair. They spend most of their time on the forest floor. But every night female gorillas clamber high into the trees. They carry twigs and leaves to make a nest. Up in the branches, they can keep themselves and their infants safe from predators. Males remain on the ground, their brawny size scaring off any possible attack.

## BEAVER

EXPLORER NOTES: *Beaver dams enclose peaceful pools of water*

Beside a fast-flowing river, brown furry mammals are busy gnawing into the trunks of tall pine trees. Beavers use their strong front teeth to cut down trees. They drag the trunks into the running water, holding them in their mouths. Their large, flat tails help them to paddle along. Trunk by trunk, branch by branch, the beavers build dams. These constructions can stretch for hundreds of metres. They form a protected space where the beavers can forage safely for food and raise their families.

## NAKED MOLE RAT

EXPLORER NOTES: *Live together in networks of tunnels*

Naked mole rats are wrinkly, hairless mammals with long teeth. They carve out tunnels beneath the soil of the East African savannah. In these underground spaces, they live together as a colony. Like ants in their nest, each mole rat has a job to do. Some dig out plants with their strong gnashers to feed the colony. Others look after the queen, a single female mole rat whose offspring will grow up into the next generation.

## HUMAN

EXPLORER NOTES: *Transforms the environment to suit its needs*

Humans are mammals that live all over the Earth. Like gorillas, beavers and even naked mole rats, they shape their own habitats too. They build towns and cities and create safe places where they can flourish. However, as cities grow, human habitats spread into the countryside. Forests and grasslands have to be cleared away, which damages the natural environment. Perhaps future humans will learn to live more responsibly with nature.

EAGLE-EYED LIBRARY

*AFRICAN ELEPHANTS*

*'Mums and dads, brothers and sisters — family is an important part of mammal life.*
*Almost all mammals give birth to live young, which first develop inside their bodies.*
*After their offspring are born, parents show their young the secrets of survival. In dense forests,*
*chimpanzee parents teach their children how to find food. On snowy mountains, snow leopard cubs*
*learn to hunt. Here are my notes on mammals' families ...'*

## The first days of life

Baby birds hatch from eggs their parents lay in nests. Tadpoles hatch from spawn frogs produce underwater. But most mammals give birth to young which are already fully formed. Whether whale or elephant, cat or gorilla, nearly all mammals first develop inside a special part of their mother's body. This unique organ is called a 'womb'. In the womb, a baby mammal absorbs healthy nutrients from the food its mother eats. The womb also protects it from the dangerous world outside. A mother leopard can pad through the jungle safely carrying her young in her womb. A mother gorilla can climb up to her bed in the tree-tops, carrying her unborn young safely up with her, protected inside her body. Then, when the time is right, a fully formed baby mammal is born.

But some mammals break the rules. The platypus is a furry mammal with webbed feet and a duck-billed snout. Its bill is incredibly sensitive. Platypus use it to detect the slightest movement of insects underwater. When the time comes to reproduce, female platypus lay eggs outside of their bodies. Then they snuggle around them for ten days until the eggs hatch.

*Most mammals first develop in a womb*

*Tigers learn key survival skills from their mothers*

## Learning from your elders

If they are to survive in the wild, infant mammals have a lot to learn as they grow up. Chimpanzees love the sweet taste of honey, but there's a knack to getting hold of the sticky liquid. Poking cautiously into a beehive, adult chimpanzees draw out strands of honey with sticks and twigs. They take care to avoid a painful sting. Young chimpanzees sit and watch as their parents suck the delicious honey from their twigs. Soon they will learn the skills themselves.

As snow falls in the Himalayan mountains, the sky turns dark as the sun sets. Hidden in the frosty crags, snow leopard cubs watch their mother as she stalks her prey. She spots a goat and pounces. The cubs quietly observe her every move. Over time the cubs will grow into deadly hunters themselves, thanks to the lessons they learned from their prowling mother.

## One of the family

Out on the African plains, elephants gather around a watering hole to drink. All the members of an elephant herd are closely related, forming a family. They travel together and help each other out as they journey on. The wilderness can be a dangerous place. If a mother is killed by a lion or a poacher, another elephant will adopt her orphaned baby. The new guardian will look after it like their own. Family is just as important for elephants as it is for humans.

Other mammals grow up in huge communities. Outside a cave, bats flit between dark trees before disappearing inside. The cave contains thousands of other bats, all roosting together. Here they are safely hidden from predators. They wait for night to fall before leaving to hunt for insects.

Most mammals, however, live in smaller groups. Tiger cubs are raised by their mothers. They may never even meet their fathers. Mothers keep a close watch over their cubs, and teach them the skills they need for life in the jungle.

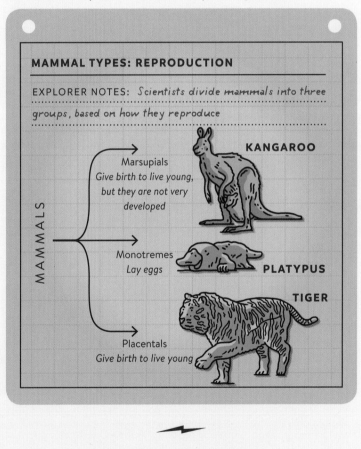

**MAMMAL TYPES: REPRODUCTION**

EXPLORER NOTES: *Scientists divide mammals into three groups, based on how they reproduce*

MAMMALS

Marsupials
*Give birth to live young, but they are not very developed*

**KANGAROO**

Monotremes
*Lay eggs*

**PLATYPUS**

Placentals
*Give birth to live young*

**TIGER**

*'Turn the page to discover different types of mammal families!'*

## Brothers and sisters

The moon shines over a snowy forest. Hidden in their den, young wolf puppies nestle together in the cold. Their soft fur keeps them warm. Their mother watches over her cubs making sure no hungry bears are prowling around outside. The forest is a dangerous place and perhaps not all the infant wolves will survive, but for now they are safe, tucked away out of sight.

## Stand up straight!

Not only do giraffes have long necks, they also have very long legs. This helps them run at great speeds across the African savannah. But before you can run, you need to learn how to walk. Mother giraffes help their babies make those first wobbly steps and stand up straight. They nuzzle them gently as they keep tumbling over. Soon they get the hang of it. Now they can start exploring their habitat on their own, placing one sturdy hoof in the ground after the other.

## Big families

Some mammals have huge families. Rabbits can give birth to over a dozen babies at once. That's a lot of baby rabbits to look after. Newborn rabbits are mostly helpless. Their eyes stay closed after they're born, so they can't see their environment. They don't yet have any fur, so they can't keep warm. Their parents protect them in a den underground called a 'warren'. A rabbit's warren can have lots of rooms and tunnels where the family hide. The baby rabbits stay safe and warm inside until they're old enough to venture out into the big world outside.

## Piggyback ride

Many mammals are constantly on the move, looking for food or water to drink. Lots of mammals take their young with them as they travel, to keep them safe. Baby monkeys cling to their mothers' backs as their parents climb up trees. Baby anteaters use their stubby claws to hang on to their parent's backs. They cling on tight as their parents search through the undergrowth to guzzle up bugs.

# 6. HUMAN MAMMALS

HUMANS

*'We've explored whales and dolphins, beavers and bats, camels and caribou, and many more mammals besides. But there's one other mammal species we might have forgotten about — human beings! Where do human beings come from? How long have humans lived on Earth? How did the early humans survive?'*

## The apes that stood up

Human beings are mammals too. They are apes, just like chimpanzees, gorillas and orang-utans. Apes first evolved 25 million years ago in Africa. The first apes spent most of their time up in the trees, climbing from branch to branch. But one group of apes clambered down from the tree-tops and started to live on the jungle floor instead. These ancient apes included the first humans.

If you look closely, you can see a lot of similarities between these ancient apes and modern humans. One example is *Australopithecus*. This ape was an ancient type of human. It had

*Human hands are adapted for grasping objects*

similar hands to a modern human. It walked upright on two feet, rather than on all fours like most other mammals. Walking upright might be the reason why humans began to invent complicated tools. Their hands were free to pick up stones or twigs from the forest floor. Soon, they started to experiment with what they found. Sharp sticks would have helped them hunt animals. Hitting stones together produced sparks which would make a fire. Modern humans continue to invent tools too. From computers to space rockets, humans make new tools to solve life's problems.

## A brief history of Homo sapiens

Many different species of humans lived in the prehistoric past. The Neanderthals were a group of humans that developed after *Australopithecus*. They had low brows and hairy bodies. But they mostly looked similar to modern humans. They walked everywhere on two feet too. Our own species, *Homo sapiens*, lived alongside Neanderthals, 150,000 years ago. Neanderthals eventually died out. Our own species continued to thrive however. Over thousands of years, our species spread across the Earth. Today, it can be found in every corner of the globe.

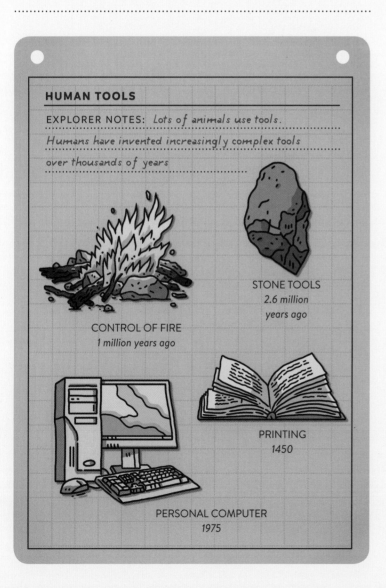

**HUMAN TOOLS**

EXPLORER NOTES: *Lots of animals use tools.*
*Humans have invented increasingly complex tools over thousands of years*

STONE TOOLS
2.6 million
years ago

CONTROL OF FIRE
1 million years ago

PRINTING
1450

PERSONAL COMPUTER
1975

*Neanderthals coexisted with our own species*

## Brains, fires and words

Wolves have sharp teeth that help them catch their prey. Buffalo have strong horns that help them fight off predators. The early humans relied on something else to help them survive – their brains. With their brains they were able to invent new things and make new discoveries. When they discovered how to control fire, their lives must have changed entirely. Night no longer trapped them in darkness. Cold weather no longer meant they would freeze. Gathered around a burning fire, early humans cooked the animals they caught, making their meat safer to eat. The early humans may also have begun to speak, using words to communicate with each other. They could pass on information about their own experiences of how the world works.

*'Turn the page to discover some of the greatest mammal discoveries of all time!'*

## Who are the great mammal Explorers?

There are many ways to explore mammals. Some explorers look at ancient mammal paintings. Others speak to mammals using a common language. Some look after ancient mammals in museums. Others think about how mammals should be treated. Here are my notes on some incredible discoveries about the lives of mammals …

### MARCEL RAVIDAT : 1922–1995

EXPLORER NOTES: *Discovered prehistoric paintings of mammals*

Marcel was a French teenager who discovered one of the greatest collections of mammal paintings ever found. Exploring the woods near his home one day, he came across a cave full of colourful images. On the cave walls were prehistoric paintings. Some of them showed giant woolly rhinoceroses, running deer and fierce bears. He returned with some friends to work out what the mysterious paintings were. Experts later investigated the discovery. They found out the pictures were made by humans 20,000 years ago.

### DOROTHEA BATE : 1878–1951

EXPLORER NOTES: *Uncovered the remains of ancient mammals*

No dangerous environment could get in the way of this fossil hunter. Travelling far from her base at the Natural History Museum in London, where she was one of the first female scientists, Dorothea Bate explored the hidden caves and sheer cliffs of Mediterranean islands. She also braved political trouble and war in her hunt for fossils. Many of her discoveries, from dwarf elephants and hippos to giant dormice, remain important today.

### BEATRIX GARDNER, ALLEN GARDNER & WASHOE : 1933–1995

EXPLORER NOTES: *Showed chimps can understand human language*

What if humans could speak to other mammals? What if we could find out what they're really thinking? Beatrix and Allen Gardner wanted to know if chimpanzees could communicate like humans. They taught some chimps to 'talk' using sign language. A chimp called Washoe even began to invent her own words. By linking up the sign for 'water' with the sign for 'bird', Washoe invented a new word for a swan – a 'water bird'.

### ROGER PAYNE : b. 1935

EXPLORER NOTES: *Recorded the songs of whales*

In 1967, Roger Payne went diving in the Pacific Ocean. He wanted to explore the way whales communicate. Roger was astonished by what he heard. It was the most beautiful thing he had experienced in the natural world. The huge whales made sounds that repeated, like the chorus in a song. Roger realised the humpbacks were singing to each other. His recordings of their songs have inspired people to protect whales.

### REBECCA BANASIAK : b. 1974

EXPLORER NOTES: *Has an unusual method for cleaning mammal fossils*

Rebecca Banasiak is in charge of the mammal section at a museum in the USA. Making sure the museum's mammal skeletons are in good condition is very important. When new skeletons are delivered, they need to be cleaned. Rebecca uses a special cleaning method. She puts the bones into a glass tank full of beetles. The beetles eat all the dirt on the bones. The skeletons are left completely ready for display!

### PAOLA CAVALIERI : b. 1950

EXPLORER NOTES: *Thinks humans should treat mammals with more respect*

Paola Cavalieri is a philosopher with an unusual interest. She studies the way human beings understand animals. What's the difference between humans and animals? Should humans treat animals in the same way as they treat other humans? She believes we should respect apes as much as we respect human life. And she believes humans need to think again about the way we treat other animals. After all, her readers are animals too – that includes me and you!

# 7. MAMMALS TODAY

EAGLE-EYED LIBRARY

BLUE WHALE

'Mammals first appeared on Earth over 200 million years ago. At first, they were mostly tiny, mouse-like animals. But over millions of years, new types of mammal developed. Today, mammals can be found in almost every environment across the world. Let's look at some of the most massive, the most mysterious and the most magnificent mammals on Earth...'

## Massive mammals

Standing at 4 metres tall, the African bush elephant is the largest animal on land. It uses its sharp tusks for digging or for defending itself. An elephant's trunk, however, is another story entirely. It can be used for almost anything. Elephants can use them for throwing heavy branches out of the way or plucking tiny seeds from a bush to eat. Elephants use their trunks to drink. They use them to spray water over their backs when it's hot. They 'hug' each other with their trunks, showing each other signs of affection. Family is important to elephants. If two elephants have been separated, they can remember each other

Elephants' trunks have many uses

when they reunite. They even get excited to see each other again!

African elephants might be gigantic, but another mammal species grows even larger. Blue whales live in the deep seas, from the Atlantic to the Indian Ocean. They grow to lengths of a whopping 30 metres. These whales are the largest animals ever to exist! They feed on tiny underwater creatures called krill. Whales sing as they swim through the dark waters of the deep oceans. Other whales can hear their songs even if they are thousands of kilometres away.

## Magnificent mammals

Mammals are survivors. In the Sahara Desert, the temperature reaches 40°C. In such conditions the sand is too hot to walk on. But the desert fox can stand the intense heat. The desert fox is a small dog-like mammal, with soft, hairy feet that protect it from the burning ground. It can use its paws to dig into the gritty sand and make a burrow to hide in, away from the blasting sun.

In the freezing cold of the Himalayan mountains, mountain goats bounce from rock to rock with perfect precision. They have padded hooves, which gives them extra grip when they spring from ledge to ledge. Their climbing skills help them escape agile snow leopards which hunt them among the rocky ravines.

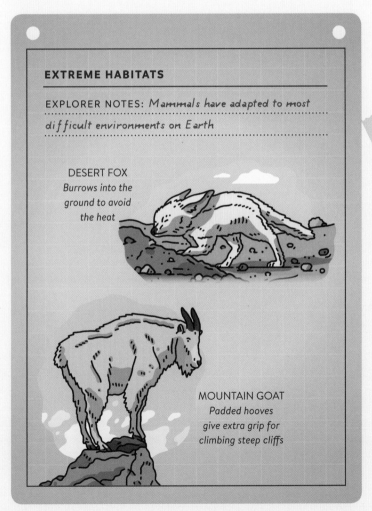

**EXTREME HABITATS**

EXPLORER NOTES: *Mammals have adapted to most difficult environments on Earth*

**DESERT FOX**
*Burrows into the ground to avoid the heat*

**MOUNTAIN GOAT**
*Padded hooves give extra grip for climbing steep cliffs*

*Lemurs are only found on the island of Madagascar*

## Mysterious mammals

Night falls over the Madagascan rainforest. A dark two-legged mammal creeps through the trees. Its body is covered in wiry black fur. Its huge yellow eyes make it look like a ghoul. But no need to fear – the creature is a kind of lemur called an aye-aye. Lemurs are primates, like apes and monkeys. Aye-ayes use their unusually long middle fingers to find the insects they like to eat. An aye-aye will tap on a tree with the first finger. If it hears a dull thump, it knows there's a bug inside. First it uses its powerful teeth to rip open a hole in the wood. Then it pokes in carefully with its third finger, hoping to snag the insect hidden inside the bark. Insects are one of an aye-aye's favourite foods.

*'Turn the page to discover some of the incredible ways humans and mammals live together!'*

## Farming

In prehistoric times, ancient humans hunted mammals for their meat. They used spears and arrows to catch deer and wild goats. They fed on the mammals' meat or used the mammals' fur for clothing. More than 10,000 years ago, some ancient tribes realised that keeping hold of the wild mammals and caring for them was easier than chasing after them all the time. These tribes were some of the first farmers. Humans still farm mammals, including cows, pigs and sheep, today.

## Companions

Attracted by the scent of a successful hunt, wolves may have visited the campsites of ancient humans. The hungry wolves hoped they could get a scrap to eat. Perhaps these ancient humans began to look after the wolves that were tamer and friendly. Some might have even kept them as part of the tribe. Over time, these tamer and more friendly wolves became dogs. Dogs have been human companions ever since. For thousands of years, they have been pets, security or helpers for sick or disabled people. Some people could not imagine their lives without the help of their dogs.

## Religion and Culture

In many human cultures, religion helps people understand how the world works. Mammals often play an important role in religious beliefs, past and present. Hindus have a special respect for cows. They are even allowed to wander into shops and along busy roads in Indian cities. In Ancient Egypt, cats were sacred. This might be because they protected the Egyptians' cities from rats and mice. For Muslims and Jews, certain mammals cannot be eaten. The Christian religion even begins with a baby being born in a stable, where a farmer kept his cattle.

## Transportation

Before the invention of cars or trains, humans relied on animals to help them get around. They used particular mammal species. People rode between towns and villages on horses. Donkeys were used to help carry heavy loads. Even today, camels carry people across deserts. Humans rely on elephants to carry heavy tree trunks. Cars and trucks are common ways of getting around. But many cultures still rely on mammals to get from *a* to *b*.

# AGENT OSPREY'S MAMMAL PUZZLERS

We've explored the whole history of how mammals developed and live on Earth – from the earliest mouse-like mammals to the great variety of species today. Now it's time to see if you've been paying attention! Let's run through some questions to check we've got our facts straight...

## WHAT'S THAT MAMMAL?

My field notes and sketches have got muddled up in my Explorer rucksack and now they're all over the place. Can you help match each snapshot of a mammal to its species?

*1.* HORSE

*2.* ELEPHANT

*3.* LION

*4.* HOUSE CAT

*5.* PIG

*6.* TIGER

# THE MAMMAL QUIZ
### Read my report carefully to discover the answers.

**1. WHAT ARE MAMMALS?**
Which of these are not common features of mammals?

*a)* Warm blood
*b)* Fur coats
*c)* Excellent vision

**2. THE FIRST MAMMALS**
*Hadrocodium*, an early mammal, was the size of which of the following?

*a)* An acorn
*b)* A bicycle wheel
*c)* A double-decker bus

**3. ANCIENT MAMMALS**
What is thought to be the largest ever meat-eating mammal?
*a)* A sabre-toothed cat
*b)* A giant bear
*c)* An enormous ground sloth

**4. MAMMAL HOMES**
Why do camels have humps?

*a)* For storing food
*b)* For storing water
*c)* For humans to ride on

**5. MAMMAL FAMILIES**
Where do most baby mammals first develop?

*a)* In a rocky den
*b)* In their mother's womb
*c)* In their mother's belly pouch

**6. HUMAN MAMMALS**
Why did early humans begin to cook their food?

*a)* To improve the taste
*b)* To remove dangerous diseases
*c)* To preserve it for eating later

**7. MAMMALS TODAY**
What are the largest mammals on Earth today?

*a)* Blue whales
*b)* Siberian tigers
*c)* African elephants

# GLOSSARY

**ADAPTATION**
A feature of a living thing which helps it survive in its environment. Also, the process of living things changing to survive more easily in an environment.

**BLUBBER**
A thick, fatty later beneath the skin of most marine mammals, such as whales and walruses. Blubber keeps them warm in cold water.

**ECHOLOCATION**
A method used by some mammals, such as bats and dolphins, to find their way and catch prey. The mammal sends out a high-pitched sound, and then listens for the echo.

**FOSSIL**
The remains of a living thing preserved for millions of years. Fossils are useful for learning about ancient living things.

**HABITAT**
The environment where a creature lives. A creature's habitat could be as small as a leaf or as large as an ocean.

**MARSUPIAL**
Mammals that give birth to live young which have not yet developed very much. Many of them then continue to develop in a pouch. Includes kangaroos and koalas.

**MONOTREME**
Mammals that lay eggs, rather than giving birth to live young. Monotremes are very rare, and include platypuses.

**PLACENTAL**
Mammal group which gives birth to live, well-developed young. Includes most mammals today, including humans.

**PREDATOR**
Any animal which hunts other animals as a food source.

**SYNAPSID**
Ancient animal group. Includes the distant ancestors of all mammals today, as well as the mammals themselves.

**TOOL**
An object that can be used to alter the environment or to make a task easier. Many animals use simple tools, but human tools are the most complex.

**WOMB**
Special organ in female placental mammals. The young are able to develop safely inside their mother's womb.

EAGLE EYED APPROVED

# THE TEAM

**NICK FORSHAW**
Nick is a writer based in Berlin. He holds a BA and MA in film and literary studies. He hopes his readers love learning about the world and all the things in it just as much as he does.

**WILLIAM EXLEY**
William is an illustrator living in Southeast London. When not illustrating Explorer books, he also works on comics. Will's favourite ancient mammal is *Gigantopithecus*.

**AQUILA NON CAPIT MUSCAS**
THE EAGLE DOES NOT CATCH FLIES

MISSION 01 — **DINOSAURS!**

MISSION 02 — **BUGS!**

MISSION 03 — **PLANTS!**

MISSION 04 — **MAMMALS!**

**EXPLORER MISSIONS**
Find out more at whatonearthbooks.com

# INDEX

Entries in *italic* can be found in the Timeline.

SECRET INDEX AGENT

# THE MAMMALS TIMELINE

We've made it back to the library! Time to look over our findings. Unfold the epic Timeline that traces the history of mammals, showcasing 100 different species. Can you find a mammal with a shell of bony plates? A rhinoceros with an enormous horn? Or a rodent as tall as a human?

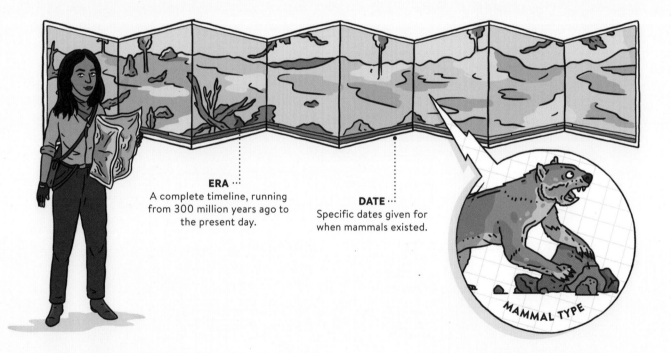

**ERA**
A complete timeline, running from 300 million years ago to the present day.

**DATE**
Specific dates given for when mammals existed.

MAMMAL TYPE

# MAMMALS!

**VOLATICOTHERIUM**
A thin skin between the front and back legs of this mammal acted like a parachute. This skin helped it glide through its forest habitat in a similar way to a flying squirrel.

**EOMAIA**
An ancestor of the placental mammals, which include most mammals today. This group develop for a long period inside a womb. This means they are less vulnerable when they are born.

**K-T EXTINCTION**
Volcanic eruptions and a meteorite impact 66 million years ago wiped out many plants and animals. The dinosaurs died out completely. Mammals manage to survive. They go on to find success in Earth's new environment.

**DIMETRODON**
This sail-backed reptile was a distant relative of mammals. Its group would give rise to our ancestors. *Dimetrodon* might have been able to control its body temperature – an adaptation seen in most mammals today.

**DINODONTOSAURUS**
This tusked plant-eater looked like a cross between a lizard and a hippo. Its group died out as the climate became drier and other creatures, including dinosaurs, were better adapted to survive.

**RUGOSODON**
One of the first of a mammal group called multituberculates. This group had a lifestyle similar to rodents today. They were successful for over 130 million years. But 35 million years ago they vanished entirely.

**FRUITAFOSSOR**
Like armadillos today, this mammal used its strong arms and long claws to dig up the nests of insects such as termites. Then it would gorge on the bugs inside.

**TEINOLOPHOS**
This unusual mammal is the earliest known monotreme. This group lays eggs like reptiles rather than giving birth to live young. Today only platypus and echidnas are left in this group.

**DIDELPHODON**
This 1m-long mammal had teeth and feet that suggest it spent lots of its life in the water, like an otter.

**CASTOROCAUDA**
This bulky, half-metre-long mammal lived some of its life in the water. Webbed feet and a flat tail like a beaver's made it an agile swimmer.

**PLACERIAS**
This sturdy plant-eater used its enormous tusks to dig through soil. When it unearthed a plant to eat, it snipped its stem with its beak.

**MEGACONUS**
Incredibly, fossils of *Megaconus* have preserved its fur coat. Fur developed before true mammals did, so some of their close ancestors already had fur coats.

**OLIGOKYPHUS**
This small, furry animal was a close relative of mammals – it may have even produced milk for its young. Its teeth were adapted for gnawing at plants, like a rat today.

**JURAMAIA**
This tiny mammal's features show it was an ancestor to placental mammals. This group nurse their young inside the womb with a special placenta organ. They include most mammals today, even us!

**ZALAMBDALESTES**
This 20cm-long mammal used its strong arms to dig holes for escaping predators or finding food.

**SINODELPHYS**
A close relative of marsupial mammals, some of which raise their young in a pouch. Today this group includes kangaroos. *Sinodelphys* used its curved claws for climbing trees to find insects to eat.

**HALDANODON**
With teeth for shearing, chewing and crushing, this mammal probably ate insects and seeds. During this period, most mammals were small, like *Haldanodon*.

**MOSCHORHINUS**
This meat-eater survived the mass extinction at the end of the Permian period. But it didn't escape unscathed. *Moschorhinus* skulls were 20 per cent smaller after the extinction. Scientists think it had adapted to survive.

**CIMEXOMYS**
Fossils of this mammal have been found at the site of an ancient dinosaur nesting colony. Its teeth show it fed on seeds and insects.

**MEGAZOSTRODON**
This small predator used its good hearing to hunt at night for bugs and reptiles. It had a warm, furry coat like a mammal. But its legs were still similar to a lizard's – it was an in-between stage.

**REPENOMAMUS**
One of the few large mammals to live alongside the dinosaurs, which were part of its diet. It ate its prey whole, including the bones!

**BIARMOSUCHUS**
This 1.5m predator was a synapsid, a group which also includes the mammals. It had large canine teeth and long legs for catching prey.

**MIACIS**
This meat-eater lived partly in trees, catc[...] mammals and raidi[...] nests for eggs. It is [...] to the Carnivora gr[...] includes cats, dogs[...]

**CRONOPIO**
Impressive canine teeth have earned this mammal the nickname 'sabre-toothed squirrel'. Despite its huge teeth, it actually ate insects.

**LYSTROSAURUS**
Fossils of this 1m-long plant-eater have been found on every continent. *Lystrosaurus* had powerful lungs to cope with lower oxygen levels. It used its tusk-like teeth to dig up roots and thick vegetation other creatures couldn't stomach.

**HADROCODIUM**
This 4cm-long animal ate insects and may have been one of the earliest mammals. It had a big brain compared to animals of similar size at this time. This probably meant it had keen senses.

*Repenomamus*

*Dimetrodon*

**ICARONYCTERIS**
To visualise its environment *Icaronycteris* made high-pitched screeches and listened to the echoes. This technique is called echolocation. Today, small bats use echolocation to hunt bugs at night. Bats are the only mammals capable of true flight.

**DARWINIUS**
A relative of lemurs. This group of monkey-like mammals are only found today on Madagascar. *Darwinius* used its long legs and large eyes to climb through the trees at night.

**BASILOSAURUS**
An early whale, a group of mammals which live in water. Scientists originally thought this marine predator was a giant reptile (its name means 'king lizard'). It lived in the ocean where it was a top predator.

**ARKTOCARA**
Like most dolphins, *Arktocara* lived at sea. Its descendants include river dolphins today. Dolphins are very intelligent and live in small groups called 'pods'. They speak to others in their pod with clicks and whistles.

**JANJUCETUS**
This whale had teeth for slicing through its prey. Its relatives today have hair-like plates called 'baleen' instead. They use these for filtering small animals from the water to eat.

**PARACERA**
The largest ever, 5.5m-t a plant-eate although it o giant stood its neck to r

**ALTIATLASIUS**
Perhaps a close relative of the first primates. Primates split early on into three main groups, lemurs, tarsiers and simians. This last group includes monkeys, apes and humans.

**SIFRHIPPUS**
This mammal was only the size of a small fox, but was actually the oldest known horse. Instead of a single hoof on each foot, it had three or four small separate ones.

**PHOLIDOCERCUS**
This small mammal had scales on its head and tail and stiff hairs on its back, similar to a hedgehog. It ate insects and fruits.

**PERUPITHECUS**
The oldest known monkey from the Americas. Monkeys originated in Africa and South America. They must have crossed the ocean to reach the Americas. They may have made the trip stranded on floating leaves and plants.

**HYAENODOIN**
This meat-eater had powerful jaws and teeth for biting prey. Bite marks from *Hyaenodon*'s teeth in fossils show it was a top predator.

**PTILODUS**
This tiny mammal ate insects and seeds. It had three types of teeth for chopping and grinding its food.

**PLESIADAPIS**
A relative of the primates, which include humans and monkeys. It still had side-facing eyes and a snout like a rodent. *Plesiadapis* had grinding teeth for chewing fruit and soft plants.

**DAEODON**
These 'hell-pigs' grew 2m-tall and probably ate both plants and meat. The tusk-like growths on *Daeodon*'s head may have protected the vulnerable parts of its head. Their group did not survive until today.

**AMPHICYON**
These muscular were related to wolves and foxes were top predata North America a

**EOMANIS**
A pangolin is a mammal with a coat of hard scales. The scales offer extra defence against predators. *Eomanis*'s legs, tail and underside were bare and unprotected.

**PEZOSIREN**
This mammal was amphibious and one of the first sirenians. Today, this group, which includes the bulky manatees, lives entirely in the water. Manatees have kept their toenails though from their time on dry land!

**ARSINOITHERIUM**
With two enormous horns, this plant-eater looked similar to a rhinoceros. But, in fact, it was related to elephants. It probably used its horns for showing off to mates or fighting rivals.

**MEGISTOTHERIUM**
One of the largest ever meat-eaters, growing 3.5m long. It probably scavenged for food when it could, instead of hunting. Meat-eaters, such as lions, often find food in a similar way today.

**MESOHIPPUS**
This horse had a large middle hoof. It would develop into the tough single hoof found on today's horses. Long legs helped it escape predators or cover large distances looking for food.

**AMBULOCETUS**
An ancestor of whales and dolphins, *Ambulocetus* still had legs for walking. However, its special ears helped it hear in water. Plus, its snout was adapted for swallowing prey beneath the waves.

**PAKICETUS**
Although it had legs, *Pakicetus* lived some of its life in the water. Its later relatives, whales and dolphins, have fully adapted to life in water.

**RUKWAPITHECUS**
Although similar to a monkey, *Rukwapithecus* was one of the first apes. These are a group of primates without tails. Apes include gibbons, chimpanzees and humans, among others.

**NECROLESTES**
This creature had strong limbs for burrowing through soil where it ate worms and insects. Its unusual features suggest it was one of the last survivors of an ancient mammal group.

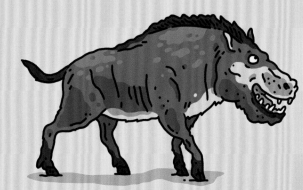

Daeodon

Icaronycteris

**DEINOTHERIUM**
A 4.5m-tall, giant relative of today's elephants. When scientists first examined its fossil, they were baffled by its backwards-facing tusks. By accident, they arranged its jaw upside down so they would point the other way!

**MEGATHERIUM**
The largest of the now-extinct plant-eating 'ground sloths'. *Megatherium* could stand 6m-tall on its hind legs to reach high-up leaves. Most giant mammals died out around 10,000 years ago at the end of the last ice age.

**CHORORAPITHECUS**
Fossils of this gorilla were found in Ethiopia. Humans and chimpanzees first developed from gorillas like these. It may even be one of our distant ancestors.

**SIVAPITHECUS**
Fossils of this 1.5m-long ape have been found across the world from Greece to China. It was related to orang-utans, which grow much larger. Today, orang-utans can only be found in Indonesia.

**ARCTOTHERIUM**
Probably the largest meat-eating mammal to ever live on land. This huge bear was 3.5m tall when standing up. It weighed a hefty 1,500kg. But its long legs made it surprisingly quick.

**PROCOPTODON**
The largest known kangaroo. It was around 2m-tall, taller than most humans.

**GIGANTOPITHECUS**
This ape's giant tooth was found in a Chinese traditional medicine shop. Scientists have identified it as probably the largest ever ape. It grew 3m-tall. That's almost twice as tall as a human!

**DESMODUS**
Most bats eat insects or fruit but these vampires bats fed on blood. After slicing a wound with their teeth, they would lap up the blood with their tongues. Today, vampire bats are much smaller than these extinct examples.

**SAHELANTHROPUS**
Perhaps the earliest hominin. This primate group includes modern humans - you and me! This ape probably already walked upright. But in other ways it looked much more similar to an ape than to today's humans.

**EPICYON**
Around 1.5m-long, *Epicyon* was the largest ever canid. This group includes dogs, wolves and foxes. It was one of the last of a unique group of dog-like predators with powerful jaws for crushing prey.

**THYLACOSMILUS**
Despite having huge canine teeth, the bite of this sabre-toothed mammal was surprisingly weak, about the same force as a house cat's! It used its teeth for puncturing the neck of its prey.

**AUSTRALOPITHECUS**
An early hominin, the group which includes humans. Like us, it walked upright on two legs. But *Australopithecus*'s brain was only a third of the size of a modern human's.

**MASTODON**
A relative of elephants and mammoths. This large hairy mammal grew to 3m-long. Its huge tusks were for competing among males. A fur coat protected it from blizzards during the last Ice Age.

**CROCERUS**
[Th]e branching horns on this [deer] were antlers, which grow [dir]ectly from the skull and are [ma]de from bone. They are [us]ually displayed by males for [co]mpeting over females.

**GIRAFFOKERYX**
This plant-eater had a shorter neck but the same furry horns as its relative, the giraffe, does today. Giraffes' long necks allow them to reach the highest leaves.

**JOSEPHOARTIGASIA**
Probably the largest ever rodent, which includes mice and rabbits. This giant grew to about 3m-long and was as tall as a human. It used its enormous teeth to cut through vegetation. They probably helped with digging and defence too.

**HOMO HABILIS**
This big-brained early human was probably one of the first to use stone tools. These tools had a sharp edge for cutting and a blunt edge for crushing.

**THYLACOLEO**
This meat-eater was a relative of kangaroos and koalas. It had powerful blade-like teeth and sharp claws for hunting.

**CAMELOPS**
Camels once roamed North America. Today they only survive in Africa and Asia. Camels are well adapted for desert conditions. Their humps store food for long journeys. Plus, they can survive months without wat[er]

**GLYPTODON**
A tough shell of bony plates protected this 1.5m-long plant-eater from predators. It wasn't enough to shield it from hungry humans, though. Our ancestors probably hunted it to extinction.

**AILURARCTOS**
This small bear was a relative of the rare giant panda. Found in South-east Asia, giant pandas only eat tough bamboo plants. *Ailurarctos* was probably still a meat-eater, though.

**MIRACINONYX**
The agile physique of this large cat was similar to a cheetah's. But it wasn't a close relative. The two species developed their similar body structures independently.

**HOMO ERECTUS**
This ancient human left Africa and spread across Asia. It was perhaps the first human to control fire. Fire allowed humans to stay warm in colder environments. Cooking food on fires also helped to prevent diseases.

**NEOTAMANDUA**
A relative of anteaters. These mammals mostly eat insects such as ants and termites which they extract from nests with their sharp claws and long tongues.

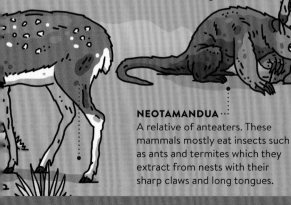

*Mastodon*

*Gigantopithecus*

**SMILODON**
This sabre-toothed cat grew over a metre tall and weighed almost a quarter of a tonne. It uses its 30cm-long canines for piercing the necks of prey after ambushing and pinning them down.

**DIRE WOLF**
These wolves had a heavier build than their close relatives, grey wolves and dogs. They hunted large animals, including bison, in packs, leaving smaller prey for the grey wolves.

**ELASMOTHERIUM**
This elephant-sized rhinoceros had a huge 1.5m-long horn. It roamed across the snow-covered plains of central Asia. A shaggy coat helped protect it from the frozen conditions.

**WOOLLY MAMMOTH**
Mammoths were relatives of elephants. Thick coats protected them from the freezing Ice Age temperatures. The last mammoths went extinct around 4,000 years ago. Rising temperatures and human hunting probably caused them to die out.

**AYE-AYE**
A lemur, aye-ayes are only found in Madagascar. Local superstitions lead to many of them being killed. Their long, spider-like finger is used to extract insects from holes in wood.

**STELLER'S SEA COW**
This 9m-long mammal was wiped out in 1768, only decades after it was discovered. It was hunted to extinction by humans for meat. Relatives include manatees.

**NARWHAL**
The tusk-like spike on this whale's head is really a giant tooth! In the Middle Ages, merchants sold it as 'unicorn horn'. The spiral of the tusk always twists in an anticlockwise direction.

**BLUE WHALE**
The biggest animals ever to have lived. Blue whales can grow 30m-long and weigh 150 tonnes. That's about 10 double-decker buses! Spread out across the oceans, blue whales communicate over many miles with deep, pulsating calls.

**MEGALOCEROS**
Depictions of this huge deer in cave art have allowed us to learn more about them. A hump across its shoulders helped to support its 4m-wide antlers.

**MACRAUCHENIA**
This strange South American mammal had a long, trunk-like nose. It was so unusual, it confused scientists for decades after its discovery.

**HORSE**
Horses were domesticated by humans living in the grassy plains north of the Black Sea. They were first used for pulling carts and chariots. Later, humans learned to ride horses.

**PLATYPUS**
These strange creatures have duck-bills and webbed feet. They also lay eggs and produce venom. They might not seem like mammals, but they are! Platypus are 'electroreceptive'. This means they can find their prey by identifying their electric signals.

**HOUSE CAT**
Cats protect humans from pests, such as snakes and rats, and subsequently disease and hunger. In Ancient Egypt, people even worshipped them. The centre of this worship was in the city of Bubastis. The cat-headed goddess Bastet was believed to protect the city.

**SIBERIAN TIGER**
The largest living cats. These tigers can reach 4m-long. Different varieties of tigers live in habitats from frozen taiga to lush rainforest. They are always top predators. Tiger mums teach their cubs to be stealthy nocturnal hunters.

**NEANDERTHAL**
These humans lived and interbred with our own species, but are now extinct. Their stout bodies kept heat in better in the cold regions where they lived. Neanderthals are some of the earliest humans known to produce art and bury their dead.

**HOMO SAPIENS**
Modern humans. They used stone, antler and bone to make tools. They even made some of the earliest art, including the first known sculptures. These humans spread across the globe, transforming environments to suit their lifestyles – they are us!

**TASMANIAN TIGER**
This wolf-like mammal was not a true tiger – it was more closely related to kangaroos! It was once a top predator in Australia and New Guinea. By the 1930s it was extinct in the wild. The last known Tasmanian tiger died in captivity in 1936.

**AUROCH**
Over time, humans bred these horned mammals into European cows. Today, cows are used for meat and milk, and for leather clothing. In some places, they are even used as currency.

**COW**
There are around 1.5 billion cows in the world today. This huge number is a source of greenhouse gases, which cause dangerous climate change.

**CHIMPANZEE**
These apes are our closest living relatives among other animals. Chimps share about 99 per cent of our DNA. They are highly intelligent and can use simple tools. Like us, they live together in complex societies.

**AFRICAN ELEPHANT**
The world's largest land mammal. Elephants grow 4m-tall and live in groups of up to 100 individuals. Elephants are threatened by poachers hunting them for their tusks.

**WOLVES AND DOGS**
Intelligent 2m-long pack-hunters. Wolves can run at 60km/h. Humans tamed wolves and bred them into today's dogs. This process is called 'domestication'.

**WILD PIG**
These mammals root through the undergrowth with their tusks in search of food. Pigs have been domesticated for meat, but eating them is taboo in many cultures.

**NAKED MOLE RAT**
Like ants, these hairless rodents build underground colonies around a single queen. They are the only mammals known to live like this. They use their large teeth to gnaw on plant roots.

**NORTH AMERICAN LION**
This 2.5m-long lion was probably the biggest ever. Lions were once found worldwide but today only live in Africa and India. Lions are unusual among cats for living and hunting in packs.

**CAVE BEAR**
Today, bears find caves to stay warm in during winter. These cave bears lived in them year round. Fossils show they fought with lions and humans, who depicted the bears in cave art.

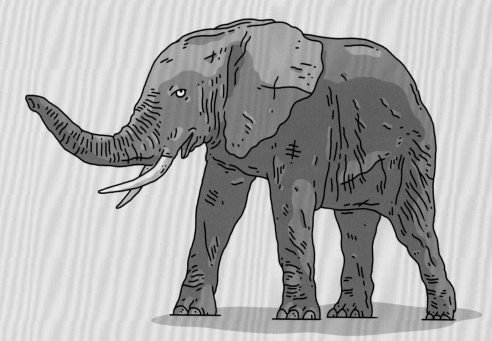

African elephant

Macrauchenia